sustainable FARMING

**HOW CAN WE
SAVE OUR
WORLD?**

sustainable FARMING

Carol Ballard

ARCTURUS

This edition first published by Arcturus Publishing
Distributed by Black Rabbit Books
123 South Broad Street
Mankato
Minnesota MN 56001

Printed in the United States

Series concept: Alex Woolf
Editor and picture researcher: Cath Senker
Designer: Phipps Design

Library of Congress Cataloging-in-Publication Data
Ballard, Carol.
  Sustainable farming / Carol Ballard.
      p. cm. -- (How can we save our world?)
  Includes index.
  ISBN 978-1-84837-290-0 (hardcover)
  1. Sustainable agriculture. I. Title. II. Series: How can we save our world?

  S494.5.S86B35 2010
  630--dc22
                              2009000620

Picture Credits
Corbis: *cover* (Andy Aitchison), 7 (Andy Aitchison), 8 (Reuters), 10 (Bettmann),
15 (Bob Sacha), 17 (Anthony Bannister/Gallo Images), 33 (Lindsay Hebberd), 37 (Andy
Aitchison), 41 (Gideon Mendel), 42 (Owen Franken); EASI-Images: 22 (Rob Bowden),
23 (Chris Fairclough), 24 (Chris Fairclough), 28 (Rob Bowden), 29 (Chris Fairclough),
34 (Chris Fairclough), 38 (Chris Fairclough), 40 (Rob Bowden) Getty Images: 16 (David
Silverman), 19 (Gianluigi Guercia/AFP), 21 (Noel Hendrickson), 25 (Christopher Furlong);
Metropolitan Council, Minneapolis-Saint Paul, Minnesota, 2006: 13; Rex Features: 9 (Sipa
Press) Science Photo Library: 26 (Françoise Sauze), 27 (Dennis Barnes/Agstockusa),
31 (Ton Kinsbergen) Shutterstock: 11 (PHOTO 888), 12 (Leonid Shcheglov), 18 (Antonio
V. Oquias), 20 (Thor Jorgen Udvang); Topfoto: 32 (Macduff Everton/The Image Works)

Artwork on pages 6 and 39 by Phipps Design

Cover picture
A farmer tends to the pineapple crop on an organic farm in Uganda, East Africa.

Credits
Every attempt has been made to clear copyright. Should there be any inadvertent
omission please apply to the publisher for rectification.

CONTENTS

The Need for Sustainable Farming

Farming, or agriculture, is the oldest and most important industry in the world. Nearly all our food comes from farming, as do the raw materials for many products such as clothing and shelter. As the global population grows, the amount of food we need also increases.

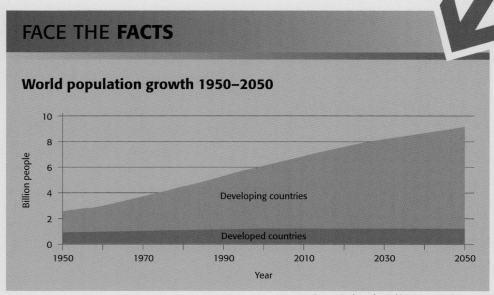

FACE THE **FACTS**

World population growth 1950–2050

This graph shows the rate at which the human population has grown since 1950 and the expected future growth. The figures for 2010 on are projections.

Source: EarthTrends, World Resources Institute, 2006

Sustainable farming

Farming uses several important resources, such as land, water, and human energy. Sustainable farming is an approach, first suggested in the 1980s, that tries to ensure that each resource is used wisely and carefully. The National

Sustainable Agriculture Information Service (ATTRA) in the US, defines sustainable agriculture as a system "that produces abundant food without depleting the earth's resources or polluting its environment."

There are five basic concepts in sustainable farming:

Produce enough food

Sustainable farming aims to ensure that the choices of farming methods, crops, and livestock are made carefully so that there will be enough food to feed the growing global population.

Avoid environmental damage

If the environment is damaged by farming, it may be unable to support food production and other human needs in the future. It is possible to farm using methods that do not cause environmental damage, and this is a key aim of sustainable farming.

Use resources efficiently

Without the efficient use of resources, it may not be possible to produce enough food for everybody. Sustainable farming aims to maximize the use of every resource, with as little waste as possible.

Be economically viable

Farming is an industry, and like all industries, it collapses if it spends more money than it makes. For long-term success, a farm must be economically viable, and sustainable farming aims to ensure that this is the case.

This mother and her son in Uganda have been trained in sustainable-farming methods. Since they began to use this approach, the quality and yield of their crops has increased.

Enhance rural life

Many people live in the countryside, and the way the land is managed has a direct impact on their lives. Sustainable farming aims to ensure that rural communities do not suffer from the agricultural activities around them. It also aims to provide employment and income for these communities.

Why is farming not sustainable now?

Ideally, farmers who plant crops or raise livestock would take into account the key concepts of sustainable farming. However, we do not live in an ideal world, and much of global agriculture is not carried out in a sustainable manner. There are several reasons for this, including

- Ignorance: many farmers have not heard of sustainable farming or do not know how to follow its principles.
- Lack of resources: many farmers have no access to the tools, seeds, and other resources needed for sustainable farming.
- Personal circumstances: individual farmers may have more regard for their own needs and those of their families than for the greater good.
- Economics: in some instances, sustainable farming may cost more and give smaller profits than other farming methods.

Problems for agriculture

Even if farmers wish to manage their land in a sustainable manner, environmental, political, and economic circumstances often prevent it. In developed countries, few people are short of food. Most have access to fresh food and clean water and can afford to buy what they need. In developing countries, though, the situation is very different. Harvests can be ruined by environmental disasters such as droughts and floods, leading to famine and

In 2001 in North Korea, the rice crop on which many people depend for food was devastated by floods. This created a serious food shortage. Here, farmers are salvaging as much rice as they can.

Sugarcane, which will be used to make biofuels, is being harvested by large machinery on this farm in Brazil. Biofuels can be used to replace gasoline in vehicles, producing far lower carbon dioxide emissions. However, biofuel crops are grown on land that could otherwise be used to grow food.

starvation for many people. In other countries, war or severe economic problems can disrupt agricultural production, causing food shortages.

While people in many developing countries suffer from food shortages, the developed world is experiencing an energy shortage. To address this difficulty, large areas of land traditionally used for food production, for example, in Brazil, are now utilized for growing biofuels. This makes the food shortages even worse. Urban development, which involves the building of houses, roads, industry, and other structures, is also taking agricultural land. The increasing global population puts agriculture under even more pressure as the demand for food increases.

PERSPECTIVE

Hunger in our world

"It is not an efficient way of meeting human needs if one billion people starve while another billion have excess. It would be more efficient to distribute resources so that at least vital needs were met everywhere. Otherwise, for example, if kids are starving somewhere, dad goes out to slash and burn the rain forest to feed them—and so would I if my kids were dying. And this kind of destruction is everyone's problem, because we live in the same ecosphere."

Karl-Henrik Robert, Swedish environmentalist

CHAPTER 1

The History of Farming

The earliest people lived on the food they found in the world around them. Many were nomadic, hunting animals and searching for berries, fruits, and nuts to satisfy their requirements. In a process that took thousands of years, agriculture slowly developed.

About 13,000 years ago, people began to sow seeds and harvest crops. They also raised animals for meat and milk. These early farmers lived in families or small groups, producing just enough food for their own immediate needs. Societies and cultures slowly developed over thousands of years, and communities gradually became more settled. As villages, towns, and cities spread, many people no longer produced their own food but were able to buy it or acquire it in exchange for other goods.

This photograph, taken in Oregon in 1880, shows an early combine harvester being used in a wheat field. Notice how many men and horses are needed to operate it!

Agricultural machinery

During the nineteenth century, advances in technology in Europe and the US led to the introduction of agricultural machinery. The first tractors powered by steam engines were introduced during the 1880s. By the

10

early twentieth century, these were superseded by gasoline-engine models. Other items of equipment that could be attached to tractors were invented, such as threshing machines and harvesters. The use of such machinery increased rapidly throughout the developed world, allowing farmers to cultivate more land and reducing the amount of work required.

Chemical fertilizers

As the workload decreased, agricultural land became more productive. Yields of some crops even doubled mostly because of the introduction of mixed chemical fertilizers, which produced healthier crops and increased yields. The first mixed chemical fertilizer, superphosphate, became available commercially in 1849. During the 1890s, an average of 1.8 million tons (1.6 million metric tons) of chemical fertilizers were used each year in the US. This figure had grown to 6.8 million tons (6.2 million metric tons) each year by the 1920s, peaking at more than 47 million tons (43 million metric tons) each year by the early 1980s.

FACE THE **FACTS**

This table shows the reduction in labor hours needed to produce 100 bushels (about 25 kilograms) of corn in the US between 1850 and 1987.

Year	Number of labor hours
1850	75–90
1890	40–50
1930	15–20
1945	10–14
1987	Less than 3

Source: United States Department of Agriculture

Here, a farmer is spraying a pesticide onto his rice crop.

Herbicides and pesticides

Other agricultural chemicals, such as herbicides and pesticides, were also developed. Herbicides kill weeds that grow alongside crops and compete with them for nutrients, light, and water, while pesticides are used mostly to kill insects. The use of herbicides and pesticides increased rapidly after their first introduction in the 1940s.

Single-crop farms

During the second half of the twentieth century, many mixed farms that reared livestock and grew crops were gradually replaced by single-crop farms. Instead of changing the use of each field every year—for example, switching between livestock pasture, crops, and resting—farmers grew one crop year after year on the same land. They removed hedges around small fields to create huge fields so that they could operate farm machinery more efficiently. Although there are still many traditional farms, especially in the developing world, much of the world's food is produced using these modern farming methods.

Storage and transportation

As farming grew more efficient, improved methods were developed for packaging and sorting the vast amount of food produced. These methods, such as canning and freezing, made it possible to store food for longer periods. Advances in transportation, such as refrigerated trucks and improved road and rail networks, enabled farmers, packagers, and retailers to transport food over greater distances than before. They could also move food quickly from one country to another by air. People were no longer reliant just on the foods grown in their area.

A vast field of corn being harvested by modern machinery. Corn is commonly grown on large single-crop farms.

Exporting food

During the second half of the twentieth century, the improvements in transportation made it easier for farmers in developing countries to export their produce to developed countries. There was a financial incentive to switch from growing traditional, life-supporting crops to producing foods that suited the export markets. Much of the profit, though, went to individuals and companies from the developed world rather than to the local farmers.

Changing land use

During the same period, as the world population rose dramatically, land was needed not only for growing food but also for housing, transportation systems, and industries. In many countries, for example, Canada and the US, landowners sold agricultural land for development, reducing the amount of land available for food production.

Toward the end of the twentieth century, governments and environmental groups around the world became increasingly aware that many human activities, including some agricultural practices, were having a bad effect on the natural environment. They tried to conserve plants and wildlife by creating areas such as nature reserves and national parks. Environmental protection as well as human needs now competed with agriculture for land.

FACE THE **FACTS**

These two aerial views taken in the metropolitan region of the Twin Cities of Minneapolis and St. Paul, Minnesota, illustrate the loss of agricultural land (green) between 2000 and 2005 as urban development (gray) takes place.

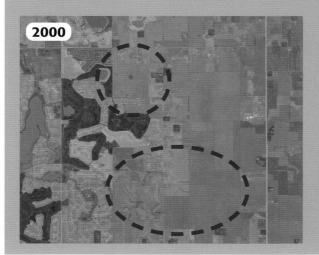

Source: Metropolitan Council, 2008

CHAPTER 2

Environmental Problems

Agriculture uses the natural resources of land, soil, air, and water. There are direct links between a farm and the environment around it. These links mean that farming practices affect not just the farmland and the food it produces but also the immediate and the wider environment.

Eutrophication

Chemicals such as fertilizers, pesticides, and herbicides pollute the land and any water that drains from it. The water carries them away, and aquatic plants and animals take them up. Different chemicals have different effects. For example, nitrates and phosphates from fertilizers increase the growth of algae in water. When the algae die, bacteria help them to decompose (break down). This process uses oxygen, and so the amount of oxygen in the water decreases. Other plants die owing to a lack of oxygen, and eventually fish and other aquatic creatures die too. The process is called eutrophication, and its effects are seen in many places around the world, including the "Dead Zone" of the Gulf of Mexico. Here, excess nutrients from agricultural land in the Mississippi and Atchafalaya river basins pour into the ocean, along with other pollutants. Nothing can live in the oxygen-depleted water.

Biomagnification

Agricultural chemicals can also build up within the bodies of aquatic organisms and in other creatures that eat them. The bigger the organism, the greater the concentration of the chemical. This is called biomagnification. A biologist named

Rachel Carson first suggested the theory in the early 1960s. Evidence from Clear Water Lake in California supported her ideas. A chemical called DDT (Dichloro-Diphenyl-Trichloroethane) was used worldwide in the 1950s as an agricultural pesticide. Soon after DDT had been used to control gnats on Clear Water Lake, large numbers of fish-eating birds began to die. Their bodies contained much higher concentrations of DDT than the lake water itself.

FACE THE **FACTS**

Farmers use herbicides and pesticides to kill weeds and crop pests. However, these chemicals also kill other plants, insects, and larger animals. If applied carefully from a tractor, they affect only a small area. However, in many places, aerial spraying is used. An airplane applies large quantities of a chemical as it flies low over the land. It is impossible to control exactly where the chemical lands, and so large areas of non-agricultural land can also be affected.

Excess growth of algae in this lake in China has caused eutrophication and turned it green. Eutrophication harms the aquatic habitat and can also be harmful to the people who fish in the lake.

Changing diets

Additional environmental problems are caused by the change in people's diets. The only foods available to the poorest people in developing countries are basic subsistence crops such as rice and other grains, beans, pulses, and roots such as potatoes. As countries develop and a proportion of the population becomes wealthier, the richer people are eager to buy protein-rich foods such as meat. This has happened in India and China since the 1980s. However, it takes far more land to raise animals for meat than to grow subsistence crops. An acre of beans, peas, or lentils can produce 10 times as much protein as an acre used for meat production. This means that a given area of agricultural land can support far more vegetarians than meat eaters.

Animals kept in confined spaces, like these hens on a farm in Israel, may produce a lot of food. However, more veterinary chemicals may be needed to maintain their health than if they were allowed greater space to move freely.

Chemicals and battery farming

Livestock production itself creates environmental problems. Many livestock farmers use a variety of chemicals. They control animal pests with pesticides, treat or prevent infections with antibiotics, and use hormones to increase milk yield and improve meat quality. Like the herbicides and pesticides used in crop production, these veterinary chemicals can cause damage to the local environment.

Veterinary chemicals are heavily used in battery farming. This is when animals, such as hens, are confined in barns or other shelters. Battery farming uses only a small amount of land and produces a high volume of cheap food, but infections in the animals spread easily in the cramped conditions. To prevent this, antibiotics are routinely used.

This sand dune in South Africa was once covered in grass, but the grass was destroyed by overgrazing. A solitary goat forages among the remaining plants.

Animals and climate change

Allowing animals to live outdoors is more humane than confining them in barns. Yet animals kept outdoors can have an adverse effect on their grazing land. Keeping a large number of animals on a single piece of land can lead to overgrazing, to the point where the vegetation cannot recover. Water for the animals may be taken from local rivers or streams, resulting in damage to the aquatic environment. Livestock emit methane and also nitrous oxide and carbon dioxide, which contribute to global warming and climate change.

FACE THE **FACTS**

If farmers want to produce the most profitable agricultural products, they may select just one or two animal strains. For example, a meat producer may rear a single type of cattle that has been bred to produce excellent meat. This specialization means that older breeds are becoming rare, and many are dying out completely. Scientists are worried about the loss of these breeds because once they die out they, and their genetic material, are lost forever.

Vital water

The increasing demand for food has also created pressures on the water supply. Water is essential both for growing crops and raising livestock. This presents a problem for farmers in areas where water is scarce. They have little choice but to take the water they need from rivers and streams. If they take too much water, the water level drops, and aquatic plants and animals cannot survive.

Land clearance

Clearing native vegetation to create new agricultural land is another major environmental issue. Farmers remove hedges and other obstacles to create large fields for growing single crops. They erect farm buildings, roads, and fences to control livestock movement. These constructions change and reduce natural habitats. Fewer wild plants and animals can survive, and the environmental balance is disturbed.

Trees in this rain forest in the Philippines have been removed in order to grow crops on the land, changing the natural environment.

For example, in tropical countries such as Indonesia and Brazil, large areas of rain forest have been cleared. While some parts are cleared by logging companies to sell the wood, much is cleared for agriculture. Some land is needed for grazing cattle to meet the increased demand for meat. Other areas are cleared for growing food crops such as soybeans to feed people and livestock and palm oil for food processing.

Damaging the soil

In places where large expanses of forest are cleared for crops, a variety of problems with the soil can develop. The roots of trees help to keep the topsoil in place; without them, the rain washes away the topsoil. If the soil is irrigated but the water does not drain adequately, it retains excess water after rain. All water contains salts, which in high concentrations can damage crops and natural vegetation. Also, if the same crop is grown continually in the same fields, the soil loses its nutrients.

SUSTAINABLE TECHNOLOGIES

Human pumps

Lack of water is a major problem for poor farmers in hot countries, but this can now be solved using human energy. Small pumps powered by hand were introduced in Bangladesh during the 1980s. Now, stronger treadle pumps, powered by moving the legs up and down on the treadles in a walking movement, have been developed. These pumps can lift water for irrigating land or for collection and storage. They are being used by more than half a million farmers in several African countries, including Zambia, Niger, and Kenya. Using human energy means no fuel costs are incurred, and the access to water significantly increases the productivity of the farms.

These children in Malawi, southern Africa, are using a treadle pump to pump water from a canal. The water will be used to irrigate this cornfield.

CHAPTER 3

Farming in a Globalized World

Many trading issues surround the worldwide agricultural industry. Some issues are relevant within a single country or region, while others are international. It is the responsibility of individual governments and international organizations to ensure that everyone receives a fair deal.

This sugarcane is being transported by truck to a local factory in Thailand.

Free trade

Free trade means that goods can be bought and sold between countries without any barriers or taxes. However, in some countries a system called protectionism exists. This means that people pay a tax on imported products. These products are sold at higher prices than produce from within that country, and so people are less likely to buy them. Although the World Trade Organization promotes free trade, the US, the European Union, and many other countries frequently ignore its rulings.

Agricultural subsidies

Subsidies also affect the price of agricultural goods. In many countries, farmers receive subsidies—payments from the government—for producing food and maintaining the land. For example, farmers throughout Europe receive subsidies through the Common Agricultural Policy, which guarantees them a minimum

price for their produce. The subsidies benefit the farmers by increasing their income so they can afford to sell their produce at a lower price than would otherwise be possible. The consumer also benefits since it keeps the cost of food low. However, farmers trying to export goods to the countries where subsidies operate have to accept lower payments. Their produce will cost more than the subsidized food, and so people will not buy it. This problem often affects farmers in poor countries.

Big business

Large companies, especially supermarkets, also influence the price of farming products. Some farmers enter into contracts with supermarkets. These can affect the farmers and the market for the produce in several ways. If the supermarket insists that it will buy nothing unless the farmers agree to sell it all their produce, the farmers are totally dependent on that company. The company can decide to reduce the amount it is willing to pay, which means that the farmer loses money. Companies may work together to force prices down and reduce farmers' incomes.

PERSPECTIVE

Farm subsidies

In 2003, Jacques Diouf, Director-General of the Food and Agriculture Organization (FAO) of the United Nations, said:

"Farm subsidies in rich countries distort the global marketplace, making it in many cases almost impossible for farmers in developing countries to compete internationally."

Much of the produce we are used to seeing on our supermarket shelves has been imported from around the world.

So why do supermarkets force prices down? In addition to covering their costs, they need to make a profit for the owners. They may cut the price at which they sell produce to encourage more people to buy from them. However, the stores often do not lose any money over these discounts—they simply reduce the amount they pay the farmers who supply the produce. Sometimes farmers receive less money for their produce than it cost them to produce it.

The need for transportation

Another factor affects the price of agricultural goods in the stores: the cost of transporting it. People in the developed world have become used to buying all kinds of foods that cannot be grown in their own country. For example, fresh fruits and vegetables such as green beans and asparagus may be flown from Kenya or Peru to the US and Europe. In turn, the economies of developing countries and their farmers rely on the export trade with developed countries. Transporting agricultural produce around the world has become a big, profitable business.

Transportation problems?

With reports showing that global warming and climate change are at least partly the result of human activities, many people are beginning to examine the way they live. This includes considering the transportation of agricultural produce. Some people believe that we should eat more local food and reduce our reliance on foods that have traveled a large number of "air miles". In the short term, the economic consequences of making such a change could be disastrous for farmers and communities in the developing countries that export these foods. In the longer term, however, it might mean that they could return to growing food for their own needs and to sell locally.

This farmworker in Kenya is harvesting green beans for export to Europe. Although the beans are an important export crop, they are not eaten by local Kenyans.

Dairy cows grazing on farmland in Ireland. The low price that dairy farmers receive for the milk they produce has made it impossible for some dairy farmers to continue in business.

PERSPECTIVE

Price of milk

In some countries, such as England, supermarkets have slashed the price that shoppers pay for milk. Many dairy farmers have gone out of business since the amount they receive from supermarkets does not cover their production costs. In 2008, Gwyn Jones, the chairman of the National Dairy Board in England, said:

"We have warned before that the era of cheap food is over and discounting of liquid milk has done irreparable damage to the sustainability of dairy farming in the past, as is all too evident now with falling milk production."

In the arid climate of Spain, this farmer is irrigating his vineyard. Without irrigation, the yield from his land would be very low.

Farming costs

The market is just one factor that farmers have to consider. They also have to work out if they can make a profit after paying the bills. The costs involved in running a farm vary from place to place. For example, growing citrus fruit in a hot, dry country such as Israel incurs irrigation costs. Without irrigation, the citrus harvest would be poor and so the income would be low. The citrus farmer must judge whether the extra income from a good citrus harvest will cover the cost of the irrigation it would require.

However, in the farming industry, it is hard to control income and costs. Unpredictable events such as floods, droughts, hurricanes, or pest infestation can seriously affect crop yields. Many farms in developed countries operate at a loss and rely on government subsidies and income from other sources to survive. In many developing countries, though, this support is not available.

Giving up

If a farm is no longer economically viable, a farmer may choose to sell the land for other uses. Agricultural land has been sold in recent years for many different uses, including tourism and leisure facilities, housing, retail outlets, business premises, and industrial buildings. For example, a report in 2001 showed that in Ontario, Canada, more than 18 percent of good-quality agricultural land had been used for buildings or transportation systems.

Alpacas come originally from the Andes in South America. Some farmers in other countries, such as the farmer below in England, now raise alpacas, mainly for their luxurious wool.

FACE THE **FACTS**

Some farms maintain income by diversifying their activities. Farmers may cultivate different crops or adopt new livestock. For example, non-native livestock such as ostriches and llamas are now found on some English farms. In the US and Europe, farmers have begun to sell their produce directly to the public through farm stores and farmers' markets. Some farmers in rural areas such as the Yorkshire Dales in the United Kingdom (UK) have converted old farm buildings for other income-generating uses such as vacation accommodation, tearooms, offices, craft workshops, and light industries. Other farms have been established as visitor attractions, such as golf courses and wildlife sanctuaries. A study in 2004 found that 17 percent of farms in England had undertaken significant diversification projects.

CHAPTER 4

Solving the Environmental Problems

Many people suggest that we should farm without the use of chemicals. This method is called organic farming. It avoids the environmental problems that agricultural chemicals cause. Scientists are developing safer chemicals for cultivating crops and new ways of increasing yields.

Safer chemicals

Some new insecticides are based on chemicals derived from plant sources. One example is pyrethrum oil, a natural insecticide that can be extracted from the pyrethrum daisy. Pyrethrum oil effectively kills insect pests but is not toxic to animals or plants. It causes no environmental damage and has been accepted for use on organic farms.

Biological controls

Biological controls involve using a biological agent, which may be a plant or tiny creature, to control pests. For example,

Marigolds grow alongside the purple Swiss chard crop in this organic kitchen garden, an example of companion planting. Marigolds deter nematode worms that attack the chard roots.

SUSTAINABLE TECHNOLOGIES

Green manure

Nitrogen is essential for healthy growth and is often supplied as nitrates in chemical fertilizers. It can be supplied naturally too. Plants such as clover, alfafa, peas, and beans are called legumes. They have nodules on their roots that contain nitrogen-fixing bacteria. The bacteria take nitrogen from the soil and make it available to the plants, thus helping the plants to grow. Other plants cannot fix nitrogen in this way. The legume crop is left to grow and then plowed back into the land as "green manure". By doing this, the nitrogen fixed by the bacteria is available for uptake by the next crop.

Soybean plants are legumes, so they fix nitrogen. These soybean plants growing among the stubble of a harvested wheat crop will increase the nitrogen content of the soil.

leafhoppers are insect pests that suck the juices from plants and weaken them. Green lacewings can be introduced to feed on the leafhoppers and control leafhopper damage in grape crops. Planting two compatible crops together can also reduce pest damage. This is called "companion planting". For instance, growing the herb borage with tomatoes can reduce tomato worm damage, and nettles can deter potato beetles.

New crops

Some scientists see the development of genetically modified (GM) crops as the solution to problems of food shortage, malnutrition, and environmental damage. New varieties such as golden rice may contain higher levels of nutrients than traditional varieties. Some GM crops are able to tolerate difficult climate conditions such as drought and salinity, while others are resistant to herbicides and pesticides. However, despite these benefits, many environmentalists are concerned about the effects on native plants and wildlife. For example, if pollen from GM plants spread to local non-GM plants, it could change the characteristics of the seeds those plants produced and affect the next generation of local plants.

Livestock and chemicals

There are also ways to reduce the use of chemicals in livestock farming. Allowing animals greater freedom and decreasing animal density lessens the risk of an infection spreading from animal to animal. This means fewer antibiotics are needed. Use of hormones can be reduced, but this may affect the yield of milk and meat and reduce the farm's profit. However, this may be set against the higher prices that many consumers are willing to pay for chemical-free foods.

Improving animal nutrition

The feeding of livestock could be made more sustainable too. People have greatly criticized feeding animals substances they would not naturally consume—and that could make them ill. For example, cattle are herbivores but are commonly fed meat and bonemeal (MBM). In 2001, following the outbreak of the deadly cattle disease bovine spongiform encephalopathy (BSE) in the UK, MBM as cattle feed was banned throughout Europe.

On this organic farm in England, cows and their calves are allowed to graze freely on pastureland. Visitors are being shown around the farm and taught the principles of organic farming.

The majority of animals kept indoors are fed commercial feed products, which the farmer has to buy. It can be cheaper, and better for animal welfare, to allow free-range grazing on pastureland. Setting aside some land for growing animal feed crops, such as hay and grain, is also a cheaper option for many farmers.

Maintaining livestock diversity

The decreasing variety of livestock can also be addressed. As specialized farms that raise a single breed of animal have become increasingly common, many breeds have become rare and in danger of dying out. Organizations such as the UK Rare Breeds Survival Trust, the American Livestock Breeds Conservancy, and the Rare Breeds Trust of Australia work to ensure the continuing survival of rare breeds by raising public awareness of the problem and by supporting individual farmers.

SUSTAINABLE TECHNOLOGIES

Energy from livestock

The problem of disposal of animal waste can be solved by using the waste as a source of energy. The animal waste is stored in underground tanks at 100 degrees Fahrenheit (38 degrees Celsius, °C), a cow's body temperature. Bacteria digest the waste to create the gas methane, which is used to power an electricity generator. The electricity produced can power farm equipment or be sold to electricity companies. Heat from the generator can be used to heat water and farm buildings. In some cases, the electricity generated from cattle waste is even used to power milking machines!

Using animal waste as an energy source is not a new idea. This woman in Agra, India, is using dried cow dung as fuel for her cooking fire. Using cow dung as fuel is common in many countries in Asia and Africa. It is also used as a fertilizer.

Land management

Good management of agricultural land, using the methods above, can help to improve yields and increase farm income. It can improve soil quality and nutrient levels, helping to maintain the long-term viability of a farm. Good land management can have environmental benefits too, reducing soil erosion and water-associated problems of flooding and drainage and reducing the need for irrigation.

Looking after the soil

A variety of methods can be used to maintain soil structure and fertility, including

- Recycling farm nutrients instead of using chemical fertilizers
- Reducing or eliminating soil disturbance by plowing less
- Ensuring that the health of soil organisms is maintained
- Using cover crops and crop residues to maintain year-round ground cover
- Using crop rotation systems to maintain soil nutrient levels and reduce disease.

Not only do these methods improve the soil quality and increase crop yields, but they can also be less expensive than other methods. For example, a decrease in plowing means less labor is required, and recycling farm nutrients is cheaper than buying costly chemical fertilizers.

Solving water problems

Water can be a problem for farmers. Too much water causes flooding, soil erosion, and the loss of soil nutrients and organisms. Too little water threatens crop and livestock survival. In areas where annual rainfall is high and flooding is a hazard, effective drainage management is needed. In places that suffer water shortages or drought, water conservation and efficient irrigation

SUSTAINABLE TECHNOLOGIES

Zero tillage

Zero tillage is a system of growing crops without plowing the land. Plant remains are left on the land after the crop has been harvested. This helps water to soak into the soil rather than running right off. As the plant remains die, their nutrients are released back into the soil. The method has been used successfully in South America since the early 1970s. A zero tillage project in India that began in 1999 has shown significant benefits over traditional plowing methods. Here, a good rice harvest was obtained even after a year of harsh weather. In addition, rice grown under the zero tillage system proved to withstand drought better than that grown with traditional plowing.

systems are essential. Improving soil management can help to solve water problems too. For example, minimizing plowing improves the soil structure, allowing the soil to retain water for longer periods.

Drainage channels, like this one in the middle of a field, allow excess water to drain away from the surrounding agricultural land.

CHAPTER 5

Meeting People's Needs

The needs of a wide range of people must be considered when developing sustainable agricultural systems. Often, the needs of different groups conflict with one another, and so efforts must be made to achieve a fair balance between them.

Farmers

Farmers need seed for growing crops. Traditionally, farmers have kept seed from one year's harvest to plant the following year. However, some modern crop varieties produce only sterile seed, so that farmers cannot do this. Instead, they have to buy new seed each year, which increases their costs. Going back to the older crop varieties would allow them to avoid these costs.

Farmers must have the means to sell their goods to others, either within their region or country or into the international market. The higher the price farmers can obtain for their produce, the greater their income, which benefits the farmers, their families, and their communities.

By sharing seeds with a neighbor, this Mexican farmer is providing him with the means to plant a crop for the next season.

This farmland in north-eastern India belongs to the family who are working on it. Many Indian agricultural workers do not possess their own land but instead have to work on other people's land in return for low wages.

Agricultural workers

Around the world, many people work on agricultural land but do not own it or have any say in its management. For example, landless laborers in India are often paid extremely low wages, and their accommodation may be linked to their job. If they lose their job, they lose their income and their home. Sustainable agriculture needs to provide agricultural workers like these with an adequate income and living conditions.

Rural communities

In most countries, a significant proportion of the population lives in rural communities and has some involvement in agriculture. These communities often lack basic amenities. They may not have a clean water supply, adequate sanitation, or central electricity. There may be no local schools, or those that exist may be poorly equipped. Inadequate transportation facilities can make it difficult to sell produce as well as limit access to medical treatment and other important services. Farming relies on rural communities for labor, but governments and large companies frequently ignore the needs of these communities.

FACE THE **FACTS**

The unequal distribution of land causes serious problems in many developing countries. One such country is South Africa, where the government has made attempts to redress the imbalance. In 2002, for example, 700 landless families were given land in Limpopo Province. Farms were set up and by 2008 were operating profitably, improving the lives of whole communities. The government intends to continue the process of redistribution. However, this means taking land away from the existing landowners, who are likely to resist the changes.

Sheep being loaded onto a ship that will transport them from Australia to the Middle East.

Consumers

Consumers need a reliable, safe, and sufficient supply of good food. In the developed world, most people have plenty to eat, but many in developing countries do not. According to the World Food Program (WFP), enough food is produced every year to feed the entire global population. However, nearly 1 billion people do not have enough to eat, and it is estimated that a child dies every five seconds from lack of food. The challenge is to improve access to food for the poorest people, at a price they can afford.

Transporters and retailers

Transporters carry produce from its place of production to the retailer. The more food that is transported, and the longer the distance it travels, the greater the profit for the transporter. The retailers can benefit from this too, since it enables them to offer a wide range of products for sale.

Although it is not true of all retailers and transporters, most are in the food business for one main reason: to make as big a profit as they can. To achieve this, they need to lower their costs to a minimum, cutting what they pay to farmers and thus reducing farmers' incomes. They also need to sell their produce at as high a price as possible, pushing up the cost of food for the consumer. The interests of large, multinational food retailers and transporters are often therefore in conflict with the needs of farmers and consumers.

PERSPECTIVE

Global food needs

According to FAO, from 2007-08 food prices worldwide rose by 52 percent. The rising cost of food is a major cause of hunger. In 2008, Oxfam, a global organization fighting poverty and hunger stated:

"People living in poverty are highly sensitive to price hikes. Around 2.7 billion people live on less than $1.40 a day and have to spend up to 80 percent of this income on food. The rising cost of basic foods (by as much as 300 percent in some places) is pushing millions of families to the limit. Many of the world's poorest people are being forced to make choices no one should have to make: parents taking their children out of school, farmers being forced to migrate to cities to live in slums, eating less and poorer-quality food."

A sustainable balance

To be sustainable, agriculture must balance the needs of different groups of people and of the environment. Measures that help one group often harm another. For example, if a government wants to pay a subsidy to farmers, it may raise the money by increasing taxes. The farmers benefit, but at a cost to the general public. Governments and organizations such as the WFP and the FAO have to try to achieve a balance of policies that benefit most and harm fewest. There is a responsibility to future generations too, and policies should conserve the environment in a way that secures the food supply in the future.

On this farm in Uganda, East Africa (opposite), the pepper crop is being grown using sustainable organic-farming methods. The farmer receives help from Kulika, a charity that trains and supports Ugandan farmers.

Helping developing countries

Food aid is not the best way forward. Food aid is distributed in emergency situations, for example, during famines and following natural disasters such as floods, earthquakes, and hurricanes. However, although it may prevent immediate suffering from hunger, emergency aid does not help to improve the long-term food supply of a country, region, or community. The most efficient ways of achieving this are through education and support. By providing practical help such as information, materials, and equipment, the developed world can help farmers in poorer countries to maximize the yields from their land. The benefits are lasting since the farmers and their communities gradually become more prosperous. Along with an improved food supply, health and living conditions improve.

SUSTAINABLE TECHNOLOGIES

Learning about sustainable agriculture

It is important to spread knowledge about sustainable agriculture. An example of this can be seen in Vietnam. In 2006, a project funded by Denmark was set up to help 8,490 Vietnamese farming households develop sustainable agriculture. The farmers were trained in improved techniques for growing corn and vegetables. They reported a 35 to 45 percent rise in corn production, and a 50 percent increase in the yield of tomatoes, cucumbers, beans, and mustard greens. The farmers learned to form groups to discuss how to improve farming methods, which has had social benefits and strengthened bonds in the community.

Working together

To develop sustainable agriculture, new ideas and approaches must be found. It is often the case that cooperation within a group of people achieves more than any individual could alone. This is true in some agricultural situations. For example, agricultural equipment such as a tractor or water pump may be too expensive for an individual farmer to buy and operate.

However, if a group of farmers all contribute to the cost and then share in its use, they can all benefit. Working together like this is called a partnership or cooperative. People can also cooperate in other ways, such as marketing and transporting their produce jointly or expanding their range of activities. For example, the National Agricultural Cooperative Marketing Federation of India (NAFED) has more than 800 members, each of whom benefits from the cooperative. NAFED buys manure, seeds, fertilizer, and machinery in bulk on behalf of its members at a lower price than an individual farmer would pay. It also helps to market the goods produced by its members.

Adding value

Even with marketing assistance, the price paid to farmers for their produce is often extremely low and may not even cover the cost of production. However, after processing, the price of the produce may go up. Some farmers and cooperatives decide to undertake the processing themselves rather than sell their produce to others. For example, in the UK, the low price of pork at the beginning of the twenty-first century meant that many pig farmers struggled financially. By taking control of processing their meat into higher-value food products such as meat pies and sausages, some farmers were able to increase their income significantly. Dairy farmers can add value to their milk by producing products such as cheese, yogurt, and ice cream. An example of this is Yeo Valley Organic, a dairy farm in England. It began making yogurts in 1974 and now also produces cheeses, butters, and other products.

Ireland has a large dairy industry. Many dairy farmers increase their profits by using their milk to produce cheese.

Vertical farming

"Vertical farming" is a method proposed for growing food in cities. It would take place in specially designed tall buildings, which have been nicknamed "farmscrapers" or "skyfarms". Like giant greenhouses, these would be powered by energy from the wind, sun, and recycled waste. The conditions inside could be controlled to allow the production of fruit, vegetables, fish, and even livestock all year round. Chemicals would be unnecessary, environmental damage would not occur, and transportation costs would be reduced. Dickson Despommier, at Columbia University in New York, estimates that a single vertical farm could produce enough food for 50,000 people. The system could be invaluable in countries where agricultural land is already scarce, such as Dubai, Iceland, and Japan.

This illustration shows how a vertical agriculture system in a city might work.

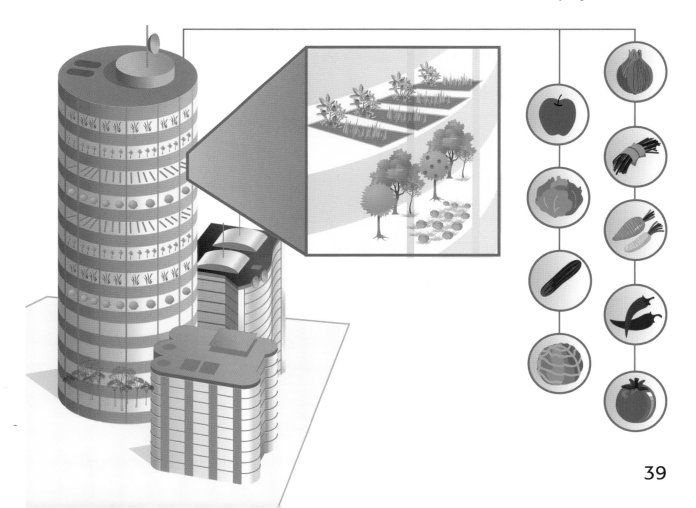

Fair trade

Another important development for sustainable agriculture has been the establishment of the fair trade movement. Many farmers in developing countries work hard and live in extremely poor conditions to produce goods that are exported to developed countries. Although the farmers receive little for their goods, the food they produce is sold for much higher prices abroad. Yet the companies that buy, transport, and sell the produce make huge profits. In the 1980s, a non-profit movement known as fair trade was developed in Europe to challenge this unfairness. It spread to other countries across the developed world. By 2008, a wide range of fair trade foods was available around the world, including coffee, tea, rice, mangoes, cocoa, sugar, wine, and nuts. The products are stocked by many leading supermarket chains and are used by restaurants, hotels, and coffee shops worldwide.

The fair trade movement works with farmers in developing countries. It sets up fair trade agreements with individual producers. Under this type of agreement, farmers receive a minimum price for their agricultural product, such as tea, coffee, or sugar. The agreement is fixed for a certain number of years to give the farmers a reliable income and some security. In turn, the farmers agree to protect the environment and to provide decent wages and conditions for farm laborers.

For example, El Ceibo is a cacao-producing cooperative in Bolivia, South America. In 1997, the cooperative began to sell cacao to the fair trade market. Half of the extra money earned is invested in the business. The rest is used to help people pay for medical treatment after accidents or to improve local school buildings. Members of the cooperative plan to expand into the production of dry fruits and coffee and maybe even promote tourism in their area.

It is not just the farmers who benefit from fair trade. When shoppers buy fair trade products, they know that their actions are helping disadvantaged groups and that farmers are protecting the environment as the foods are grown.

Fair trade products like these are available in many developed countries. Both the farmers who produced them and the environment benefit from the sale of these products.

These women are carrying baskets of coffee beans to a fair trade market in Haiti. The beans were grown by farmers who are part of a local cooperative that helps them to process and market their coffee beans.

PERSPECTIVE

Helping businesses to develop

"The guarantee of the minimum price brings stability. We, producers, are not totally subjected to the law of supply and demand. We know that we will be paid at least $69 the quintal [220 pounds, or 100 kilograms]. This guarantee makes it possible to plan long term, to invest, to develop technical support—to develop our business."

Felipe Cancari Capcha, a producer from El Ceibo Cooperative, a fair trade cacao producer in Bolivia.

The fresh produce here has been grown locally and is being sold direct to the public at a farmers' market in North Carolina.

Farmers' markets

Farmers' markets show how farmers can cooperate to help each other in developed countries. Also known as green markets, they give independent farmers the opportunity to sell their produce directly to the public. The markets benefit the farmers because they are not tied to the demands of large retailers and manufacturers. They are also good for the customers, who gain access to fresh, cheap produce. There are advantages for the environment too. Selling produce in the locality where it was grown uses less energy than transporting it over large distances. Packaging is also reduced. Farmers' markets are becoming increasingly popular with farmers, consumers, and entire communities. In the US, for example, the number rose from 1,755 in 1994 to 4,685 in 2008—a rise of more than 250 percent.

The future of agriculture

Sustainable agriculture requires careful land management and control of environmental issues such as the effects of transportation. The potential benefits offered by biotechnologies such as genetic modification must be balanced against any possible dangers. Rural communities should be supported and enabled to prosper. Above all, sustainable agriculture must strive to ensure that the food needs of the global population can be met, both now and in the future.

PERSPECTIVE

Hope for the future

At the High-Level Conference on Food Security, held in Rome, Italy, in June 2008, officials of 181 countries around the world signed a declaration that stated:

"We firmly resolve to use all means to alleviate the suffering caused by the current crisis, to stimulate food production and to increase investment in agriculture, to address obstacles to food access, and to use the planet's resources sustainably, for present and future generations. We commit to eliminating hunger and to securing food for all today and tomorrow."

This commitment, if fulfilled, will ensure that in future generations, every person will have enough to eat and the fragile environment and limited resources of our planet will be protected.

algae Very simple plants that grow in or near water.

antibiotic A medicine used to treat or prevent infections.

aquatic Growing or living in, on, or near water.

bacteria Tiny forms of life that exist in air, water, and soil and in dead and living things. Bacteria often cause disease.

biomagnification The process by which the concentration of chemicals increases with every step up the food chain.

cooperative A business that is owned and run by the people involved, who share the profits.

crop rotation The practice of planting different crops on a field each year to prevent the increase of specific pests and to improve the soil.

diversify Develop a wider range of products.

diversity A range of many people or animals that are very different from one another.

drainage Removal of excess water from land.

drought A long period with little rain.

ecosphere The atmosphere, oceans, plant, and animal life, and top part of the earth's crust.

eutrophication When chemicals in water cause a huge growth in algae, which use up the oxygen. Other plants and animals die.

fertilizer A product, often made from chemicals, that farmers may add to the soil or water to help plants grow.

genetic To do with the genes, the units in the cells of a living thing that control its physical features.

genetic modification (GM) Changing the genetic makeup of an organism.

herbicide A chemical used to kill weeds.

hormone A chemical that can increase animal growth.

insecticide A chemical used to kill insect pests.

irrigation Supplying water to an area of land through pipes or channels to help crops to grow.

livestock Farm animals such as cattle, sheep, and poultry.

malnutrition Poor health caused by a lack of food or a lack of the right type of food.

methane A gas that is emitted by some animals and can be used as a fuel.

monoculture Growing a single type of crop or livestock.

nutrient A substance that is needed to keep a living thing alive and help it to grow.

organic farming A farming method that avoids the use of man-made chemicals.

pesticide A chemical used to kill pests on crops and livestock.

pollute To add dirty or harmful substances to land, water, or air so that it is no longer pleasant or safe to use.

protectionism Taxation of goods entering a country to make them more costly than locally produced goods.

protein A natural substance found in meat, eggs, fish, and some vegetables that humans and animals need to help them to grow and stay healthy.

residue An amount of something that remains at the end of a process.

resistant Not affected by something.

retailer A person or organization that sells goods to the public.

salinity The amount of dissolved salts in water.

soil erosion The wearing away of the land through natural processes or through activities such as mining or the grazing of animals.

sterile Unable to develop into new plants or to produce offspring.

subsidy Money paid by governments to reduce the cost of producing goods so that their price can be kept low.

subsistence crop A basic crop that gives farmers and their families just enough food to live on.

sustainable Involving the use of natural products and energy in a way that does not harm the environment.

tillage The process of turning the soil by methods such as plowing or hoeing.

topsoil The layer of soil nearest to the surface of the ground.

viable Able to survive.

World Trade Organization An international organization that enforces the rules of global trade.

yield The amount produced, for example, of a crop.

Website disclaimer
Note to parents and teachers: Every effort has been made by the publishers to ensure that these websites are suitable for children, that they are of the highest educational value, and that they contain no inappropriate or offensive material. However, because of the nature of the Internet, it is impossible to guarantee that the contents of these sites will not be altered. We strongly advise that Internet access is supervised by a responsible adult.

BOOKS

Baines, John D. *Global Village: Food and Farming.* Evans Brothers, 2008.

Baines, John. *Sustainable Futures: Food for Life.* Evans Brothers, 2005.

Ballard, Carol. *Global Questions: Is Our Food Safe?* Arcturus Publishing, 2008.

Bowden, Rob. *Sustainable World: Food and Farming.* Wayland, 2007.

Casper, Julie Kerr. *Natural Resources: Agriculture.* Chelsea House Publishers, 2007.

Macfarlane, Katherine. *Our Environment: Pesticides.* KidHaven Press, 2007.

Spilsbury, Louise. *Food and Agriculture: How We Use the Land.* Raintree, 2006.

WEBSITES

http://attra.ncat.org/new_pubs/attra-pub/PDF/sustagintro.pdf?id=other
The website of ATTRA, the National Sustainable Agriculture Information Service in the US, provides information about sustainable farming.

http://www.agclassroom.org/kids/index.htm
Part of the US Department of Agriculture website, designed for children and teenagers with information, games, and activities about agriculture.

http://www.defra.gov.uk/farm/index.htm
This section of the UK Department for Environment, Food and Rural Affairs website covers all aspects of farming.

http://www.fairtrade.org.uk/
The Fair Trade Foundation website gives information about fair trade, its aims and its projects.

http://www.farmbureaukids.com/kidscorner/kidscorner.html
A website with information and activities about farming in the US.

http://www.soilassociation.org/
The website of the Soil Association covers organic and sustainable farming.

http://www.ukagriculture.com/farming_today/sustainable_agriculture.cfm
This site includes all aspects of farming in the UK.

45

INDEX

Page numbers in **BOLD** refer to illustrations.